Wildebeest Migration

By Susan H. Gray

21st Century Junior Library

CHERRY LAKE PRESS

Published in the United States of America by Cherry Lake Publishing Group
Ann Arbor, Michigan
www.cherrylakepublishing.com

Reading Adviser: Marla Conn, MS, Ed., Literacy specialist, Read-Ability, Inc.
Content Adviser: Sheila K. Schueller, Ph.D., Lecturer and Academic Program Specialist, University of Michigan School for Environment & Sustainability
Photo Credits: ©Oleg Znamenskiy/Shutterstock.com, cover, 4; ©Tony Campbell/Shutterstock.com, 6; ©GUDKOV ANDREY/Shutterstock.com, 8, 18; ©Tina Case/Shutterstock.com, 10; ©Susan Schmitz/Shutterstock.com, 12; ©AdangRuj/Shutterstock.com, 14; ©Michael Potter11/Shutterstock.com, 16; ©Mcknub/Shutterstock.com, 20

Cherry Lake Press is an imprint of Cherry Lake Publishing Group.

Library of Congress Cataloging-in-Publication Data

Names: Gray, Susan Heinrichs, author.
Title: Wildebeest migration / by Susan H. Gray.
Description: Ann Arbor, Michigan: Cherry Lake Publishing, [2021] | Series: Marvelous migrations | Includes index. | Audience: Grades 2-3
Identifiers: LCCN 2020002785 (print) | LCCN 2020002786 (ebook) | ISBN 9781534168619 (hardcover) | ISBN 9781534170292 (paperback) | ISBN 9781534172135 (pdf) | ISBN 9781534173972 (ebook)
Subjects: LCSH: Gnus—Migration—Juvenile literature.
Classification: LCC QL737.U53 G674 2021 (print) | LCC QL737.U53 (ebook) | DDC 599.64/591568—dc23
LC record available at https://lccn.loc.gov/2020002785
LC ebook record available at https://lccn.loc.gov/2020002786

Cherry Lake Publishing Group would like to acknowledge the work of the Partnership for 21st Century Learning, a Network of Battelle for Kids. Please visit http://www.battelleforkids.org/networks/p21 for more information.

Printed in the United States of America
Corporate Graphics

CONTENTS

5 The New Year

9 One Big Loop

15 Rains, Rains, and Animals of the Plains

19 What a Trip!

22 Glossary

23 Find Out More

24 Index

24 About the Author

A wildebeest is also known as a gnu.

The New Year

It's January on Africa's **Serengeti** plain. Afternoon temperatures are around 80 degrees Fahrenheit (27 degrees Celsius). Wildebeest herds arrived here a month ago.

Make a Guess!

Some animal mothers go off alone to have their babies. But wildebeest moms give birth inside the herd. What is the advantage of doing this?

On their journey, wildebeests are **prey** to waiting lions and cheetahs.

Thanks to recent rains, tasty grass is everywhere. For wildebeests, life is good.

February comes, and wildebeest moms have their calves. Throughout March, wildebeests feed and the young grow stronger. Soon, others join the group. Before long, hundreds of thousands of wildebeests are on the move. The Great **Migration** has begun.

Think!

Migration does not occur at exactly the same time every year. Wildebeests might start moving in March of one year, but in April the next. What do you think might cause such changes? Read more online or at your library to see what experts say.

A wildebeest gets its name from its wild appearance.

One Big Loop

The wildebeests have just started a long journey. The trip can stretch out to 1,000 miles (1,609 kilometers). The animals will travel in a gigantic loop through parts of

Look!

Wildebeests are closely related to cattle, sheep, and water buffalo. Look at some photographs of these animals. What features do they have in common?

Wildebeests are a tasty treat for crocodiles.

eastern Africa. The search for good grazing lands keeps them on the move. In April, May, and June, the herds keep heading north. Zebras and **gazelles** travel right along with them. Altogether, there may be 2 million animals.

Around July, they come up against their first big river. The wildebeests do not cross it just yet. July is the driest month for the entire region. Wildebeests are desperate for a drink. They bunch up in tight, noisy crowds and gather at the river's edge to satisfy their thirst. They have to be careful, though. Hungry crocodiles swim in these waters.

Thousands of wildebeests die each year attempting to cross the Mara River.

Despite the crocodiles, wildebeests must get to the other side. A few plunge in and others follow. Once they reach dry land, the animals continue their migration.

Next, they encounter the Mara River. Crossing is a real challenge. First, a few brave animals enter the rushing water. Then, others just pour in. Some take flying leaps into the river. It's so crowded they even land on each other. One herd can take hours to cross. And there are many herds.

Wildebeests will graze until sunset.

Rains, Rains, and Animals of the Plains

Around September, wildebeests reach the northern end of their migration loop. The **climate** here is wetter than the southern lands that they left. There's enough water and grass to go around.

Rain usually comes with afternoon thunderstorms. Such **downpours** feed the rivers where crocodiles and hippopotamuses swim. Rain also fills streams and ponds across the plains. These become watering

The dung beetle also helps fertilize the soil.

holes for wildebeests, zebras, cheetahs, hyenas, and wild dogs.

One creature here does not get much attention—the **dung** beetle. Dung beetles are the clean-up crew for the animal migrants. These little insects gather animal droppings and lay eggs on them. The dung balls provide food for beetle **larvae** when they hatch.

Ask Questions!

In Africa, wildebeests live to be around 20 years old. In zoos, they usually live longer. Why would wildebeests have shorter lives in the wild?

Wildebeests can run at speeds of up to 50 miles (80 km) per hour.

What a Trip!

By the end of October, wildebeest herds are turning south. For the next 2 months, they continue on. In December, they finally return to the southern Serengeti. The journey has not been an easy one. The animals faced **predators** and possible drowning.

In Africa, there are other wildebeest herds that *never* migrate. Scientists are trying to understand this. What drives some

People travel from around the world to
witness the wildebeest migration.

herds to go on such a risky trip? What makes others simply stay put?

Scientists might puzzle over this for years. As for the migrating wildebeests, the older ones have completed their 20th trip. For those born last February, it was their first big adventure. Soon, most of the herd's females will give birth. And another migration will begin soon after.

GLOSSARY

climate (KLYE-mit) the overall weather in a region

downpours (DOUN-porz) heavy rains

dung (DUNG) animal droppings

gazelles (guh-ZELZ) hoofed animals that move gracefully

larvae (LAR-vee) very young, undeveloped insects

migration (mye-GRAY-shuhn) movement of animals from one region to another and back again

predators (PRED-uh-turz) animals that hunt and eat other animals

prey (PRAY) animals that are eaten by other animals

Serengeti (sehr-un-GET-ee) a large region in northern Tanzania and southern Kenya

FIND OUT MORE

BOOKS

Carney, Elizabeth. *Great Migrations: Whales, Wildebeests, Butterflies, Elephants, and Other Amazing Animals on the Move*. Washington, DC: National Geographic Children's Books, 2010.

Fishman, Jon. *The Wildebeest's Journey*. Minneapolis, MN: Lerner Publications, 2018.

Samuelson, Benjamin. *Journey of the Wildebeests*. New York, NY: Gareth Stevens Publishing, 2018.

WEBSITES

Animal Corner—Wildebeest

https://animalcorner.co.uk/animals/wildebeest
Learn about the wildebeest's body, behavior, and sounds and the ways it helps the Serengeti's environment.

Fort Wayne Children's Zoo—Wildebeest

https://kidszoo.org/our-animals/wildebeest
Find more details about the wildebeest, including what it eats at the zoo.

INDEX

Africa, 5, 11, 17, 19

babies, 5, 7, 21

crocodiles, 10, 11, 13, 15

dung beetles, 16, 17

food, 7, 11, 14, 17

gazelles, 11
gnu. *See* wildebeest
grazing, 11, 14
Great Migration, 7

herds, 5, 11, 13, 19, 21

Mara River, 12, 13
migration
  miles traveled, 9
  when it happens, 7, 11
  where it happens, 9–14
mothers, 5, 7

predators/prey, 6, 10, 11, 13, 19

rain, 7, 15
rivers, 11, 12, 13, 15
running, 18

Serengeti plain, 5, 19

watering holes, 15, 17
wildebeests, 4, 8
  deaths, 12
  how long they live, 17
  predators/prey, 6, 10, 11, 13, 19
  running speed, 18

zebras, 11, 17

ABOUT THE AUTHOR

Susan H. Gray has a master's degree in zoology. She has written more than 170 reference books for children and especially loves writing about animals. Susan lives in Cabot, Arkansas, with her husband, Michael, and many pets.